# ANOTHER AMAZING
# TRIP AROUND THE SUN...

.......................................................................

.......................................................................

.......................................................................

.......................................................................

.......................................................................

Made in the USA
Columbia, SC
30 January 2022

55061803R00057